AF271410

A PHOT OF THE DWARF PLANET PLUTO
(PHOTO COURTESY OF NASA)

Let's Learn About Dwarf Planets
© 2018-2025 Keith Tarrier
Original Illustrations by Keith Tarrier.
Photos of planets by NASA

ISBN: 9781790196593

Tarrier Design Worldwide,
Maroochydore, Queensland, Australia
Nihonbashi-Ningyocho, Chuo-ku, Tokyo, Japan
www.keithtarrier.com

www.keithtarrier.com/books

SCAN ME

VISIT
TARRIER
BOOKS
AMAZON
USA
<--------

TARRIER BOOKS
AUSTRALIA • JAPAN

To the young minds who ask 'why?'—never stop.

LET'S LEARN ABOUT DWARF PLANETS

NASA'S NEW HORIZON AT PLUTO

WHAT IS A DWARF PLANET?

- A dwarf planet is a type of small planet, but it is not considered a full planet like Earth or Jupiter.

- The International Astronomical Union (IAU) defines a dwarf planet using these rules:

- It must orbit a star (like the Sun).

- It must have enough gravity to become a rounded shape.

- It must not be a moon (satellite) of another object.

- It has not cleared its orbit of other nearby objects.

- The only difference between a dwarf planet and a regular planet is that a dwarf planet shares its orbit with other space objects.

COMPARISON OF PLANETS TO DWARF PLANETS

Jupiter

 The Moon

Earth

Charon
Pluto

Made By Keith Tarrier

HOW MANY DWARF PLANETS ARE THERE?

• There are currently five dwarf planets officially recognized by the International Astronomical Union (IAU).

• In order of distance from the Sun, they are: Ceres, Pluto, Haumea, Makemake, and Eris.

• Other likely dwarf planets include Quaoar and Sedna, both discovered in 2002, as well as Orcus and Salacia, discovered in 2004.

• Scientists estimate there may be up to 200 dwarf planets in our solar system, especially in the region beyond Neptune.

5 RECOGNISED DWARF PLANETS

Dysonmia

Nix

S/2012 P1

S/2011 P1

Charon

Hydra

Eris　　**Pluto**　　**Makemake**

Namaka

Hi'iaka

Haumea　　**Ceres**

OTHER KNOWN DWARF PLANETS

Weywot

Vanth

2007 OR$_{10}$　　**Quaoar**　　**Orcus**

Sedna

Made By Keith Tarrier

WHERE ARE THE DWARF PLANETS?

• Most of the dwarf planets we know today are located in or beyond the Kuiper Belt.

• The Kuiper Belt is a region filled with millions of asteroids, comets, and frozen objects.

• It begins just beyond Neptune, about 30 AU from the sun, and extends far into the outer solar system.

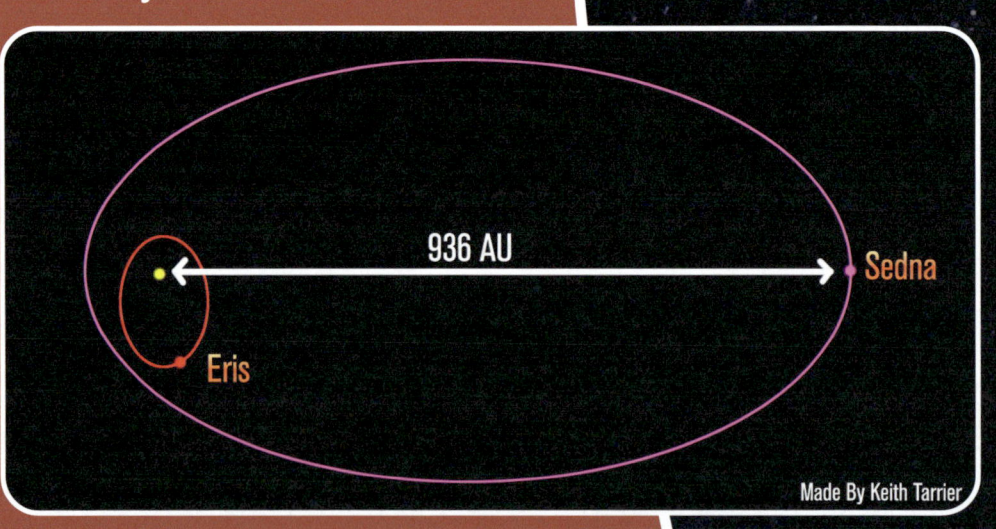

936 AU   Sedna

Eris

Made By Keith Tarrier

WHERE ARE THEY?

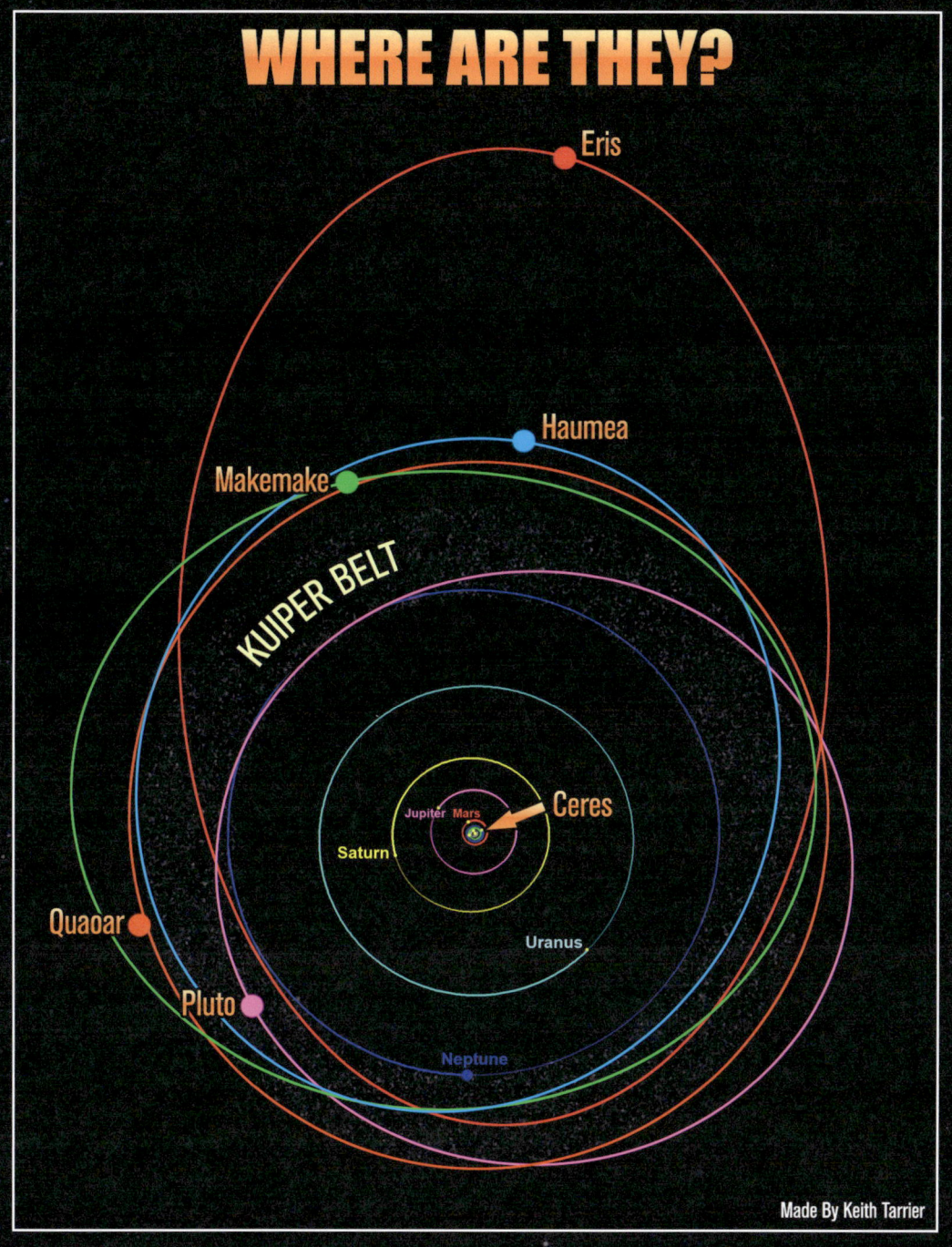

KUIPER BELT

Eris
Haumea
Makemake
Quaoar
Pluto
Ceres
Jupiter Mars
Saturn
Uranus
Neptune

Made By Keith Tarrier

CERES

• Ceres was discovered by Italian astronomer Giuseppe Piazzi on January 1, 1801 — 45 years before Neptune.

• It is the largest object in the asteroid belt between the orbits of Mars and Jupiter.

• Ceres is the closest known dwarf planet to Earth.

• The NASA spacecraft Dawn entered orbit around Ceres on March 6, 2015, and studied it in detail.

• Ceres orbits the Sun between 2.5 and 2.9 AU.

• It has a radius of about 473 km, making it the smallest recognized dwarf planet.

• Ceres has a mass about 12% that of Earth's moon.

• Scientists discovered bright spots on its surface, which are made of salty minerals — most likely a type of sodium carbonate left behind by briny water.

• Evidence suggests that Ceres may have a layer of water ice beneath its surface, and possibly even ancient underground oceans.

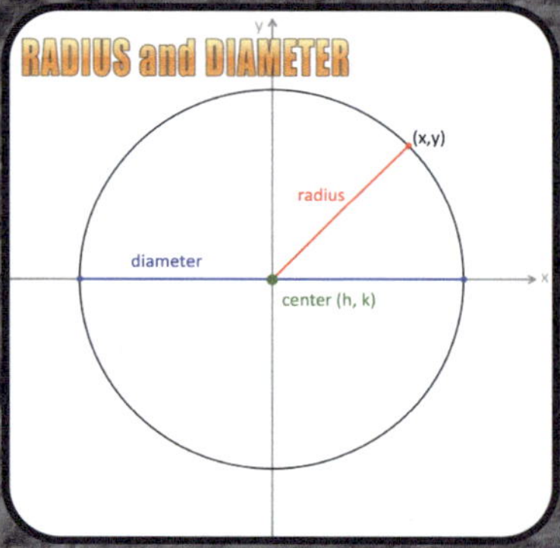

RADIUS and DIAMETER

(x,y)

radius

diameter

center (h, k)

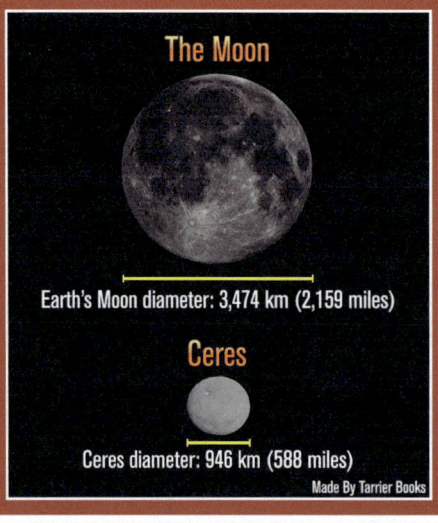

The Moon

Earth's Moon diameter: 3,474 km (2,159 miles)

Ceres

Ceres diameter: 946 km (588 miles)

Made By Tarrier Books

This is Occator Crater. It is 57 miles (92 kilometers) across and 2.5 miles (4 kilometers) deep. The crater contains the brightest area on Ceres.

DAWN SPACECRAFT

- Dawn is a space probe launched by NASA in September 2007.

- Its mission was to study two large objects in the asteroid belt: Vesta and Ceres.

- Dawn is the first spacecraft to orbit two different extraterrestrial bodies in one mission.

- It was also the first spacecraft to visit either Vesta or Ceres, and the first to orbit a dwarf planet.

- Dawn arrived at Vesta in July 2011 and studied it for over a year before continuing to Ceres.

- It entered orbit around Ceres in March 2015 and spent more than three years studying its surface.

- The spacecraft used special ion engines, which allowed it to travel long distances using very little fuel.

- Dawn's mission ended in October 2018, but it remains in orbit around Ceres as a silent observer.

Illustration By Keith Tarrier

An artist's vision of Dawn spacecraft approaching Ceres.

PLUTO

• Pluto was discovered by Clyde Tombaugh in 1930. It was originally considered to be the ninth planet from the Sun.

• The name Pluto was proposed by Venetia Burney (1918–2009), an eleven-year-old schoolgirl in Oxford, England.

• Pluto is named after the Roman god of the underworld.

• Pluto is the largest and second-most-massive known dwarf planet in the solar system, after Eris in mass.

• On July 14, 2015, the New Horizons spacecraft became the first spacecraft to fly by Pluto.

• Pluto has a radius of about 1,188 km, which is roughly 19% the size of Earth.

• You would need about 5.5 Plutos to equal the mass of Earth's moon.

• Pluto orbits between 29 and 49 AU from the Sun. For part of its orbit, it is actually closer to the Sun than Neptune.

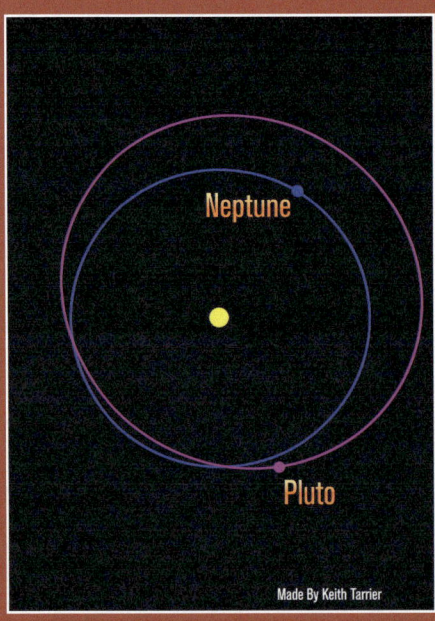

Made By Keith Tarrier

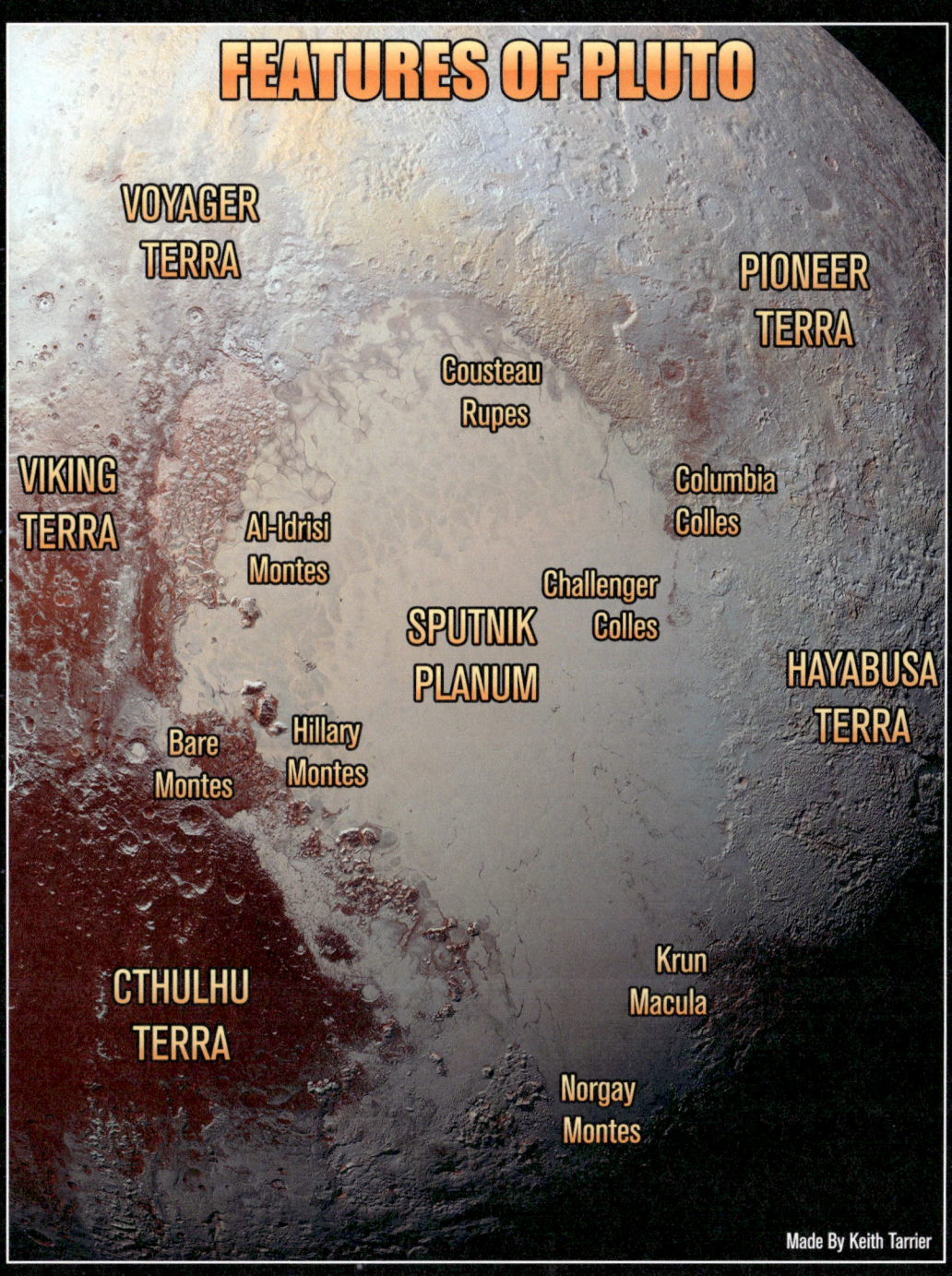

FEATURES OF PLUTO

VOYAGER TERRA

PIONEER TERRA

Cousteau Rupes

VIKING TERRA

Columbia Colles

Al-Idrisi Montes

Challenger Colles

SPUTNIK PLANUM

HAYABUSA TERRA

Bare Montes

Hillary Montes

Krun Macula

CTHULHU TERRA

Norgay Montes

Made By Keith Tarrier

PLUTO'S MOONS

- Pluto has five moons known to be larger than 1 km in diameter.

- In order of distance from Pluto, they are Charon, Styx, Nix, Kerberos, and Hydra.

- Charon is the largest of the five moons.

- Pluto–Charon is sometimes considered a double dwarf planet.

- Charon's diameter is 1,212 km, about half that of Pluto.

- Styx has a radius of 7 km, Kerberos has 12 km radius, Nix is 42 km radius, and Hydra has a 55 km radius.

- (The radius is the distance from the center of a planet to its edge.)

Charon and the Small Moons of Pluto

Styx　　　　Nix　　　　Kerberos　　　　Hydra

10 miles
10 km

Charon

CHARON

• Charon was discovered in 1978 by James Christy at the United States Naval Observatory in Washington, D.C.

• It is the largest of Pluto's five known moons.

• Charon has about half the diameter and one-eighth the mass of Pluto.

• This makes Charon unusually large compared to its parent planet — so large that Pluto and Charon are sometimes called a double dwarf planet system.

Charon (rear) and Pluto

• Unlike most moons, Charon and Pluto always show the same side to each other — they are locked in a special kind of orbit called mutual tidal locking.

WHY THE NAME CHARON?

• In the 1940 science fiction novel *Calling Captain Future*, author Edmond Hamilton named the then unknown three moons of Pluto: Charon, Styx, and Cerberus.

• James Christy decided to use the name Charon after discovering that it is a Greek mythological figure. Charon is closely associated in myth with the god Hades. The Romans identified Hades with their god Pluto.

NEW HORIZONS SPACECRAFT

- New Horizons is a NASA space probe launched in January 2006.

- Its main mission was to explore Pluto and its moons, making it the first spacecraft to visit a dwarf planet.

- On July 14, 2015, New Horizons flew past Pluto, sending back the first close-up images of Pluto, Charon, and the entire Pluto system.

- The mission revealed icy mountains, flowing nitrogen glaciers, and a large heart-shaped region on Pluto's surface.

- After Pluto, New Horizons continued deeper into the Kuiper Belt and flew past a small object called Arrokoth on January 1, 2019.

- New Horizons used a special type of power called a radioisotope thermoelectric generator (RTG), since sunlight is too weak that far from the Sun.

- The spacecraft is still sending data and exploring the outer regions of the solar system today.

New Horizons near Pluto and Charon

Illustration by Keith Tarrier

10 Years and
3 Billion Miles...
(4.8 billion km)

Earth
January 19 2006

New Horizons launches from Cape Canaveral, Florida in the USA.

Jupiter
February 28 2007

The spacecraft flies past Jupiter. The gravity of Jupiter makes it fly faster. This make the journey to Pluto three years quicker.

December 2015

The spacecraft is awakened and gets ready for its very busy flyby of Pluto and Charon.

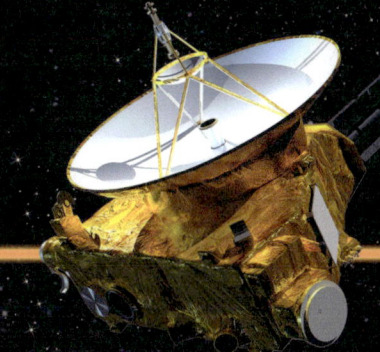

Pluto
July 14 2015

New Horizons is at its closest to Pluto. It takes many wonderful photos and gathers scientific information.

2007 - 2014

During this time the spacecraft sleeps. Once a week it sends a short message back to earth that says "I am sleeping peacefully."

Illustration By Keith Tarrier

HAUMEA

• Haumea is named after the Hawaiian goddess of childbirth and fertility.

• It was discovered in 2003 by a team led by Mike Brown at the Palomar Observatory in California.

• Haumea is one of the fastest-spinning large objects in our solar system — it completes one rotation in less than four hours.

• Its rapid spin has stretched it into an unusual ellipsoid shape, making it look like an American football or an egg.

• Haumea orbits the Sun at a distance between 34 and 54 AU.

Artist illustration of what Haumea may look like.

- Its mass is about one-third that of Pluto, and around 1/1,400 the mass of Earth.

- Haumea has a radius of about 816 km, although its stretched shape makes its dimensions vary depending on direction.

- It has two moons named Hiʻiaka and Namaka, both named after daughters of the goddess Haumea in Hawaiian mythology.

- In 2017, astronomers discovered that Haumea has a ring system — the first ring ever found around a trans-Neptunian object.

Illustration By Keith Tarrier

An artist's impression of the dwarf planet Haumea, with its two moons and thin ring system.

MAKEMAKE

• Makemake was discovered on March 31, 2005, by a team led by Michael E. Brown.

• It was the fourth object to be officially classified as a dwarf planet.

• The name comes from Makemake, the creator god in the mythology of the Rapa Nui people of Easter Island.

• Makemake is one of the largest known objects in the Kuiper Belt, with a diameter about two-thirds that of Pluto.

• It orbits the Sun at a distance between 38.5 and 52.8 AU.

• One year on Makemake takes about 310 Earth-

The Sun and inner solar system as seen from the distant dwarf planet Makemake — located over 6 billion kilometers (nearly 4 billion miles) from the Sun. A spacecraft like New Horizons would take over 16 years to reach it.
(Image Credit: Tarrier Books original illustration)

years.

• Makemake has a radius of approximately 739 km.

• It has one known moon, nicknamed MK 2, which is estimated to have a radius of about 87 km.

• Makemake rotates once every 22.5 hours, giving it a day just slightly shorter than Earth's.

• Unlike Pluto, Makemake has no atmosphere — at least none that we've been able to detect so far.

• Makemake appears very bright because its surface is covered with frozen methane, which reflects sunlight.

Illustration By Keith Tarrier

Artist's illustration of the view from Makemake, looking back toward Earth and the Sun — about 45 to 53 AU away. At that distance, sunlight takes around 6 to 7 hours to reach the surface.

ERIS

- Eris is named after the Greek goddess of strife and discord.

- It was discovered in 2005 by astronomers Mike Brown, David Rabinowitz, and Chad Trujillo.

- Eris orbits the Sun at a distance between 38 and 68 AU.

- One year on Eris takes about 558 Earth-years.

- Eris has a radius of approximately 1,163 km, making it slightly smaller than Earth's moon but more massive than Pluto.

- It has a small moon called Dysnomia, which orbits Eris once every 16 days.

- The discovery of Eris led to the redefinition of the word "planet" and the creation of the new category — dwarf planet.

Artist's illustration of a possible future mission to the distant dwarf planet Eris and its moon, Dysnomia. If a spacecraft similar to New Horizons were launched when Eris is at its closest point to Earth, the journey could still take more than 20 years to complete, depending on launch speed and trajectory.

Illustration By Keith Tarrier

OTHER DWARF PLANETS

• In addition to the five officially recognized dwarf planets, astronomers have discovered several other likely dwarf planets beyond Neptune.

• Four of the best-known candidates near the Kuiper Belt are Orcus, 2002 MS4, Salacia, and Quaoar.

• These objects orbit the Sun at distances between 39 and 67 AU and take about 250 to 300 Earth-years to complete one orbit.

• Each of them has a diameter between 850 and 1,100 km, making them large enough to possibly be rounded by gravity — one of the main

Illustration By Keith Tarrier
Artist illustration of 2002MS4.

requirements for being a dwarf planet.

• Orcus was named after a Roman god of the underworld. In mythology, Orcus punished the wicked, while Pluto ruled over the entire underworld.

• Another fascinating dwarf planet candidate is Sedna, which lies far beyond the Kuiper Belt.

• Sedna has a highly stretched orbit that takes it from about 76 AU to as far as 940 AU from the Sun.

• One year on Sedna is incredibly long — about 11,400 Earth-years!

• Sedna is one of the most distant objects ever found in our solar system and may belong to a region called the inner Oort Cloud.

• Many more distant dwarf planets may still be out there, waiting to be discovered as telescopes become more powerful and new missions explore the far reaches of space.

Illustration By Keith Tarrier
Artist illustration of Sedna.

GOBLIN

• Nicknamed "The Goblin," this small icy world was discovered around Halloween in 2015 and announced in 2018.

• Its official name is 2015 TG387, but it got its spooky nickname from the time of year it was first spotted.

• The Goblin is only about 300 km wide, much smaller than Pluto.

• It has an extremely stretched orbit, traveling from 65 AU to over 2,300 AU from the Sun.

• One year on The Goblin lasts about 44,000 Earth-years.

• Some scientists think its strange orbit may be a

clue that a much larger planet — sometimes called Planet Nine — could be hiding in the far reaches of the solar system.

Goblin

Goblin, Sedna and Eris's orbits.

Illustration By Keith Tarrier

Artist illustration of what The Goblin may look like.

THE FUTURE

- New telescopes and sky surveys are helping astronomers discover more and more objects beyond Neptune.

- Some of these distant worlds may soon be added to the official list of dwarf planets.

- The farther we look, the more we find — and the search is just beginning.

- Who knows — maybe one day, you will discover a new dwarf planet!

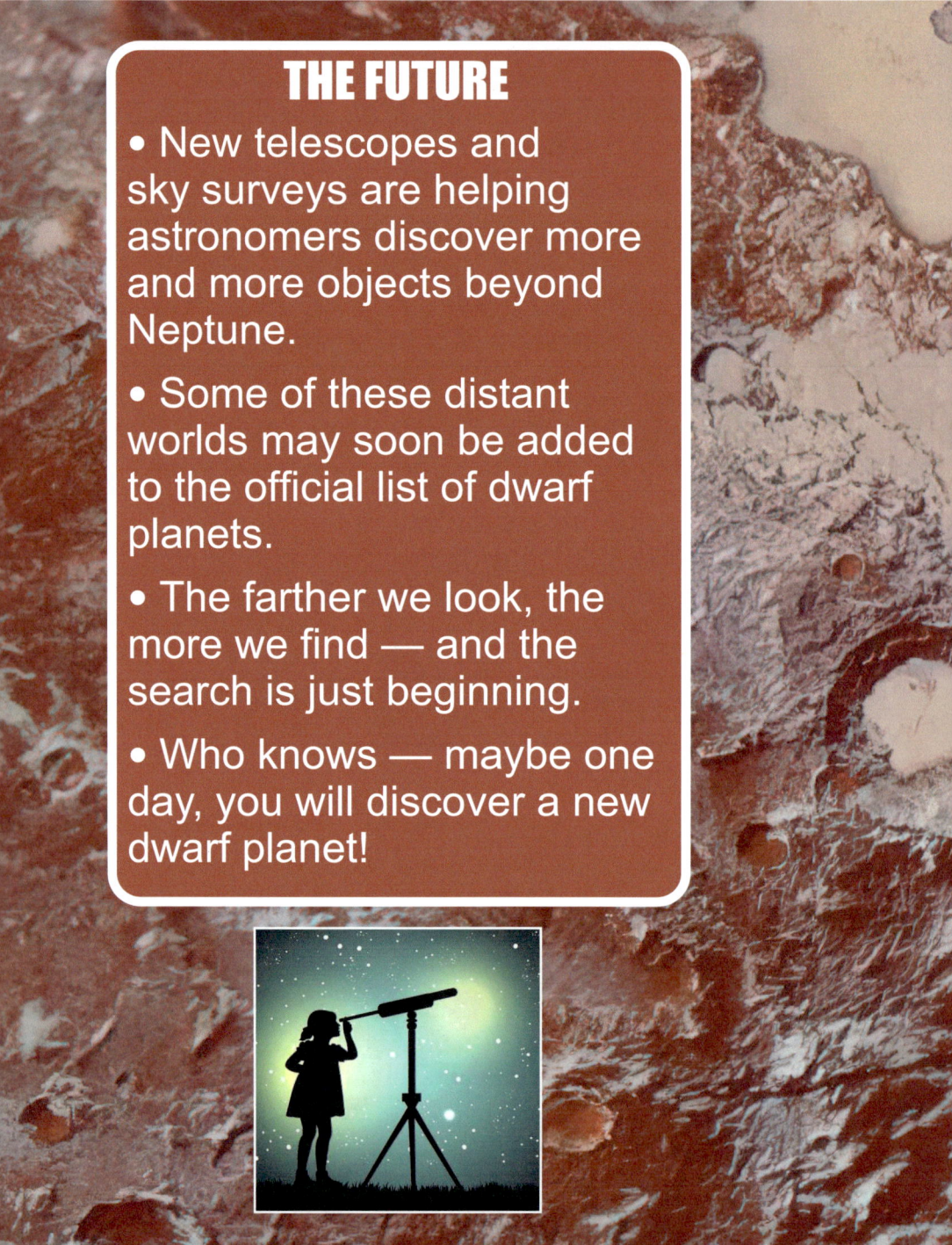

This enhanced color image of Pluto shows how different and interesting its surface is. You can see dark red areas with lots of craters, bright icy plains like Sputnik Planum with blocky mountains, and frozen methane on ridges and slopes. Some spots even have small pits caused by ice turning into gas!
(Image credit: NASA/JHUAPL/SwRI)

GLOSSARY

Astronomical Unit (AU) – A way to measure distances in space. One AU is the distance from Earth to the Sun — about 150 million kilometers (93 million miles).

Axis – An invisible line a planet or moon spins around. This spin creates day and night.

Celestial Object – A natural thing in space, like a planet, moon, asteroid, or star.

Comet – A small icy object that orbits the Sun. When it gets close to the Sun, it heats up and releases gas and dust, creating a glowing tail.

Core – The center part of a planet or moon.

Crater – A round hole or dip in the surface, made when something like a space rock hits it.

Diameter – A straight line from one side of a circle to the other, passing through the center.

Dwarf Planet – A small planet-like object that orbits the Sun, is round, but hasn't cleared its orbit of other space objects.

IAU (International Astronomical Union) – A group of scientists who decide official names for space objects like planets, moons, and asteroids.

Light-Year – The distance light travels in one year: about 9.5 trillion kilometers (or 5.9 trillion miles). It's used to measure distances between stars and galaxies.

NASA – The American space agency. They send astronauts and spacecraft to explore space.

Kuiper Belt – A huge ring of icy bodies, comets, and dwarf planets beyond Neptune.

Orbit – The path a planet or moon takes as it moves around a star or planet.

Protoplanet – A large space rock that was on its way to becoming a planet when the solar system was forming.

Radius – The distance from the center of a circle to its edge.

Rotation – The spin of a planet or moon on its axis. One full spin is one day.

Satellite – Something that orbits a planet. Natural satellites are moons. We also build satellites that orbit Earth.

Solar System – The Sun, all the planets, moons, and space objects that orbit it.

Space Probe – A robot spacecraft sent to explore space. Probes take pictures and send back information.

Tholins – Reddish, sticky substances made when sunlight hits gases like methane and nitrogen. They give Pluto and other distant objects their reddish color.

TNO (Trans-Neptunian Object) – A space object that orbits the Sun farther out than Neptune.

Illustration By Keith Terri

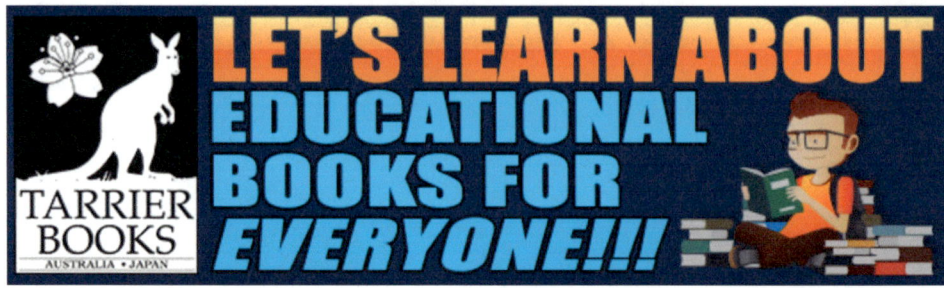
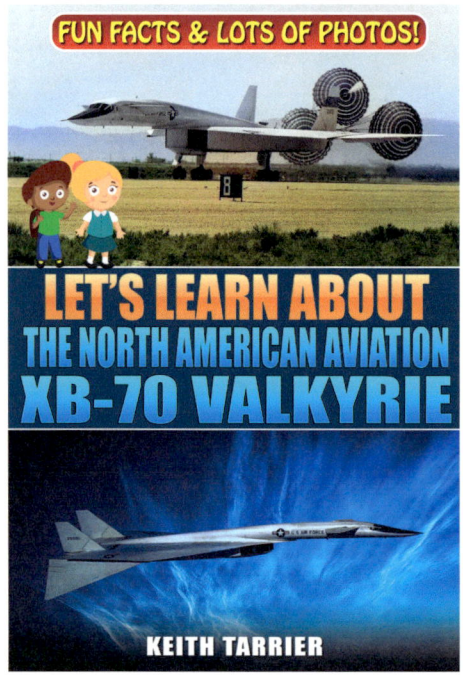
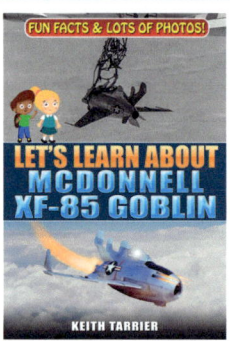
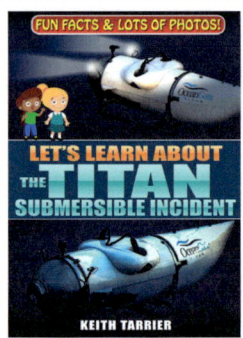
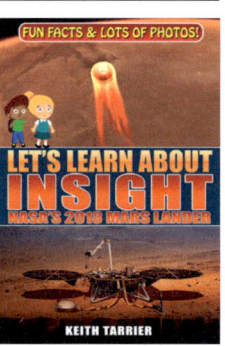

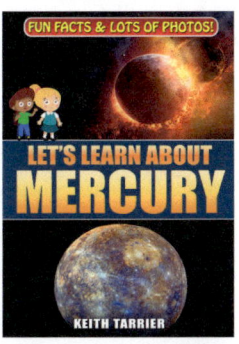

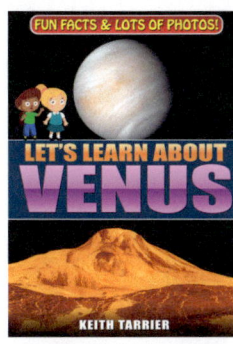

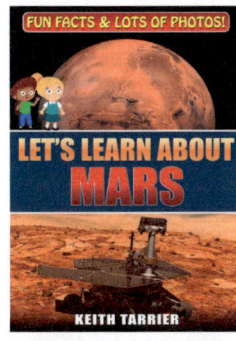

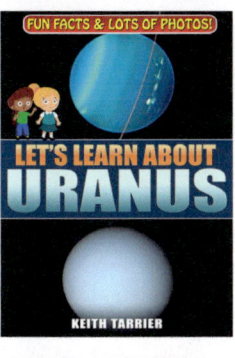

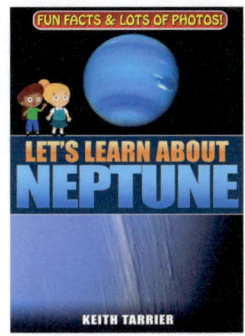

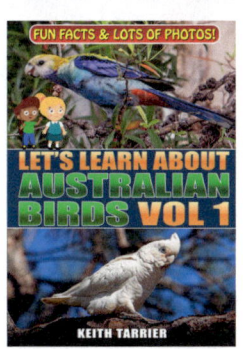

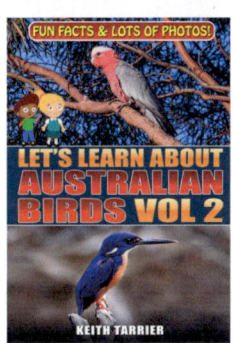

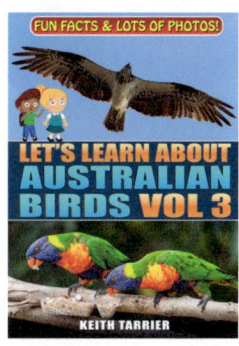

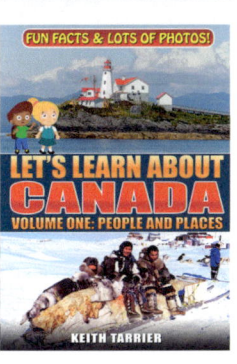

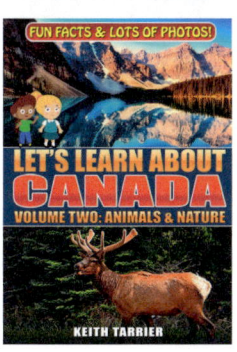

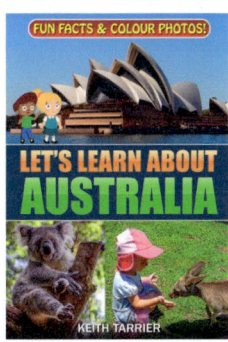

ALSO AVAILABLE FROM TARRIER BOOKS:

COLORING BOOKS!

Volumes One and Two of the *Wacky Aliens Coloring Books 1 and 2. Wacky Monsters Coloring Books 1 and 2, Wacky Creatures and Motocross Action.*

20 strange and fun illustrations in each book!

If your kids love this book, they will love *Wacky Aliens, monster and creatures* too!

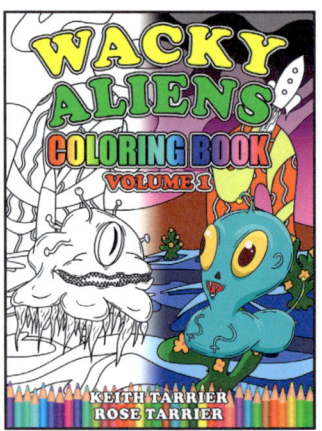
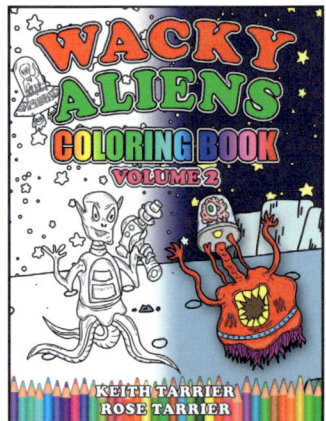
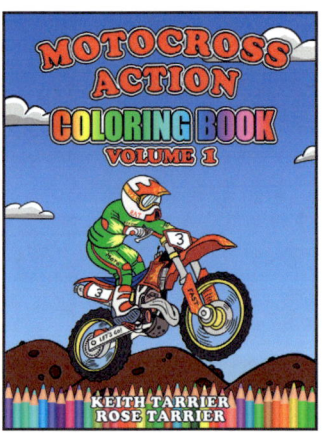
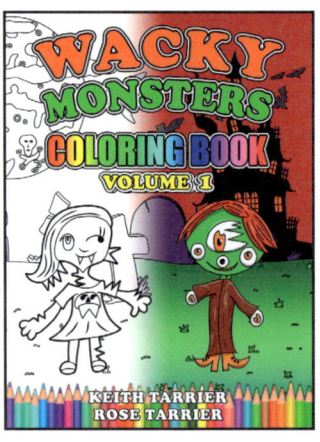
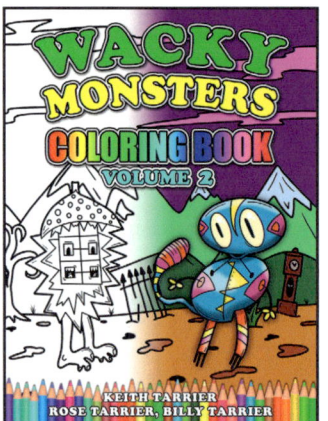
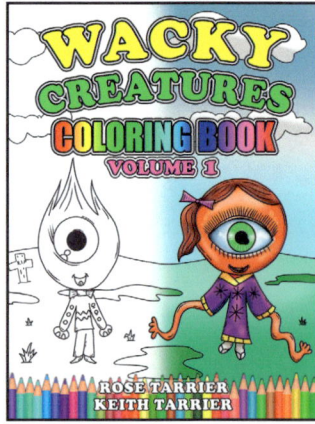

For these popular children's books and more please visit:
amazon.com/author/keith-tarrier

FUN STORYBOOKS THAT DELIGHT EVERYONE:

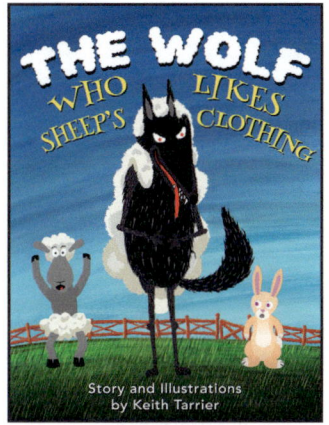
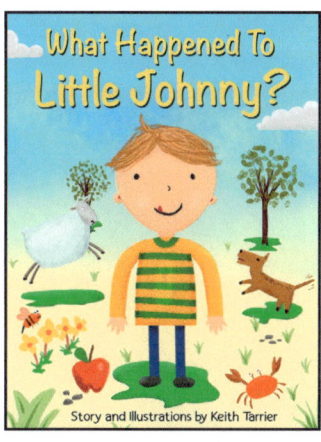
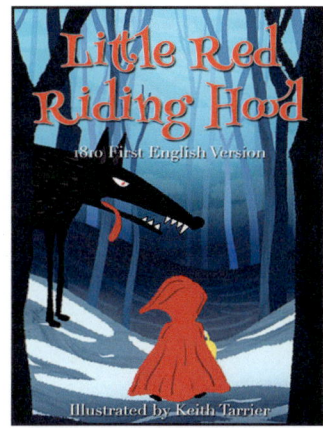

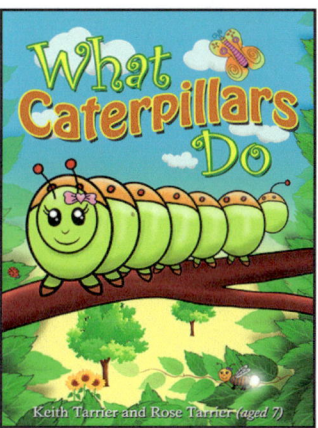
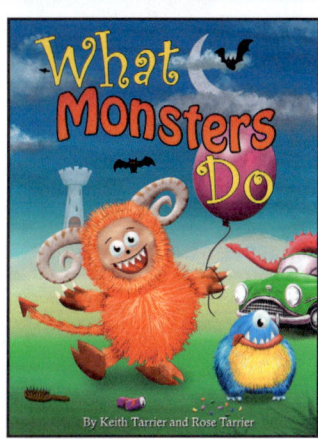

LEARN COUNTING AND COLORS:

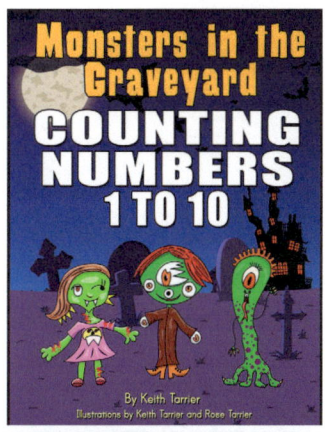
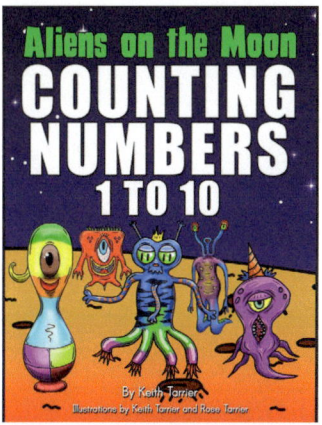
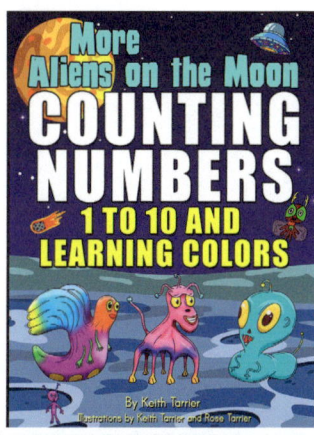

Made in the USA
Monee, IL
07 July 2026

56551416R00033